G000293467

These Cryptograms belongs to:

CRYPTOGRAM

1.

X	E	K	H		C	U		E	U	K		H	G	W	W	T

2.

B	U	M	D	B	K		P	M	K		2	7		A	X	X	D	K

3.

M	S	A	W	R	I	F	A	K	D		R	S	B	M		5

R	M	S	A	W	D

4.

A	D	I	O	Y		E	B	O	Y	P		O	Y	B

I	B	T	N	-	U	O	W	J	B	J

5.

S	B	W	L	W		J	F		N	H	U	I	J	C	W		J	I

U	Y	Y	O	W		Y	J	Y	F

CRYPTOGRAM

6.

| Q | | E | H | Z | Q | K | | W | O | W | M | Q | J | J |

| L | W | X | U | E | C | | Q | K | | T | H | K | P | W |

7.

| Y | C | R | O | H | | F | S | K | O | J | Q | | O | R | T | J | C |

| E | F | R | E | R | O | Y | K | J | | E | Y | N | J |

8.

| J | V | . | | V | P | F | O | V | T | | Q | T | | Y | S |

| P | V | U | Y | Q | S | O | U | | J | Q | S | Q | T | H | O | V |

9.

| X | S | Q | | V | Y | J | J | W | | T | S | L | | E | Z | U | J |

| 2 | 0 | 0 | | X | Y | | F | X | Y | J | | G | J | S | Y | W |

CRYPTOGRAM

10.

| C | M | B | Y | D | U | | H | N | U | I | Z | M | A | | D | J | Y |

| I | U | F | M | | J | W | | D | B | U | I | | D | I | | 1 | 9 | 3 | 5 |

--

11.

| U | T | R | | X | B | Q | U | | O | B | O | C | K | E | J |

| P | J | B | V | L | | N | C | K | N | Q | | E | J | R |

| U | C | K | F | O | Q |

--

12.

| C | | T | G | D | J | Y | | D | N | | M | C | G | S | W | | E | W |

| X | C | M | M | F | L | | C | P | | F | R | C | M | K | C | K | E | D | P |

--

13.

| W | A | | C | K | G | H | C | P | G | , | | U | M | P | Q | | F | M | K | G |

| O | W | H | | C | S | W | Z | Y | | 1 | 5 | | T | G | C | H | Q |

--

CRYPTOGRAM

14.

L	G	Y		G	Z	Q	Z	R	R	Z	D		Z	X	B	G	Z	M	Y	L

V	D	X	I		G	Z	U		1	2		X	Y	L	L	Y	O	U

--

15.

L		U	K	M		Y	L	Z		B	K	G	O

U	K	N	E	L	O	H	N	S	C	H	Z		J	C		J	H	Z

Z	H	K	N	L	U	Y

--

16.

X	B	X	Z	Q		I	T	Q		2	,	7	0	0

O	X	A	O	C	X		I	N	X		A	W		V	X	T	Z	H

I	N	D	X	T	D	X

--

CRYPTOGRAM

17. JMQS 90% JL YJCKJU

QHYJKISQK JGGIS CU

RJEQK

--

18. CDRUR LUR FIRU 1,800

GJFZJ XMROARX FV

VYRLX

--

19. OPBCKBXFL FZ GKY HYLP

BH OPFVY BP OPFVLGFVS

--

CRYPTOGRAM

20.

X	G		K	W	I		P	B	S	W	I	,		R	W	I	X	G	Y	I

C	Y		G	Y	Q		E	W	P	B		W	G	H

R	K	Y	R	Z	I

--

21.

G	S	P		L	A	R	G		B	A	B	I	M	H	W

Q	W	H	C	O		A	X		W	H	U	R	U	C	R		U	R

R	I	C	L	H	U	O

--

22.

Q	W	P	D	R	T	F	P		J	K	R	P	O	X	F	P

F	P	H	W	P	U	W	Y		U	V	W		K	I	L	O	F	P	A

L	V	R	F	K	.

--

CRYPTOGRAM

23.

V	G	V	E	H		3	0		M	V	K	S	L	Y	M		R

F	S	P	M	V		C	N	E	V		Y	S	P	A	Q	V	M		N	L

M	N	T	V

24.

G	C	N		G	J	S		G	V		V	G	J	L

W	D	N	L	O	M	W	C		U	G	J	H	N	U	G	F	E	H

G	Z	C		W		M	W	S

25.

G	F	M	T	M		P	T	M		L	U	M	T		5	0	0

I	D	W	W	M	T	M	Y	G		G	C	O	M	Q		L	W

K	P	Y	P	Y	P	Q

CRYPTOGRAM

26.

NIHTH STH 293 ASBX NZ

FSPH YISQCH RZT S

WZJJST

--

27.

FXAF AK FCQ IBHQ

NAWQI YGS B LSQNIBIF

NGTPYAKC

--

28.

LINDKN ZNQ GVKN

DVQCNC UQ VQC

NPRDUWNC LGN ZUUQ

--

CRYPTOGRAM

29.

| H | A | F | R | A | A | M | | 1 | 2 | % | - | 1 | 5 | % | | Q | O |

| F | I | A | | L | Q | L | G | E | C | F | N | Q | M | | N | V |

| E | A | O | F | - | I | C | M | J | A | J |

30.

| Z | L | W | | F | O | I | G | Z | G | W | | J | W | | R | Q | G |

| A | L | W | R | | S | L | Z | Z | X | R | G | B | | H | J | R | N | | J | O |

| R | Q | G | | X | W | F |

31.

| P | E | T | L | T | | D | L | T | | Q | C | L | T | | P | E | D | U |

| 6 | 4 | 0 | | Q | K | F | B | X | T | F | | V | U | | P | E | T |

| E | K | Q | D | U | | S | C | Y | G |

CRYPTOGRAM

32.

5 0 0 , 0 0 0   J V F T   V I   S U H

L T   Q V X H   V I   T Z C H   T H D

U A L T H U A Q F T

--

33.

U M T T V Q   C O   N F V   O V U M X E

J M O N   D O V E   J V N K I   C X

N F V   H M Q I E .

--

34.

I F C D   4 %   I B   S X S V W P   X K W

S I K F   I F   M J W V K   X A M Z X C

U Z W   U X M W

--

CRYPTOGRAM

35.

7 5 - 9 0 % | E K | D T O C Q T F

D R F N O X O Q L | H O N O G N | Q T Z

M A Z | G E | N G T Z N N

--

36.

Z J | H J | X L H P , | H J

H D L P H E L | Y L P F G J | Q H R L F

1 , 1 4 0 | Y C G J L | V H I I F

--

37.

P W T E : S U H V H Q P W T E ,

X W U U H Q P W T E ,

E S T O H Q P W T E , | N Q P

X N M N U P W T E

CRYPTOGRAM

38.

C	F		W	Q	M	U	W	Y	M		E	M	C	E	T	M

B	M	W	U		P	E	D	Z	M	U	P		O	C	U	M

R	I	W	F		R	I	M	G		Z	C		Z	M	W	R	I

--

39.

F	N	P		G	O	V	S	P		Z	P	F	Y	P	P	I

A	T	L	C		P	A	P	Z	C	T	Y	G		Q	G

S	V	U	U	P	M		F	N	P		X	U	V	Z	P	U	U	V

--

40.

Q	I	U	W	U		E	W	U

E	X	X	W	J	S	V	C	E	Q	U	R	A

7	5	,	0	0	0	,	0	0	0		I	J	W	M	U	M		V	L

Q	I	U		D	J	W	R	F

--

CRYPTOGRAM - HINTS

#		#	
1.	U => O, G => L, W => E	29.	Q => O, N => I, A => E
2.	K => S, A => M, X => O	30.	W => S, I => G, X => U
3.	A => R, M => E, D => S	31.	L => R, P => T, F => S
4.	U => H, N => T, D => O	32.	T => S, U => H, V => I
5.	O => L, I => N, N => C	33.	O => S, F => H, V => E
6.	M => B, H => U, W => E	34.	C => L, I => O, S => B
7.	E => C, C => D, K => T	35.	G => T, M => D, Z => E
8.	T => S, O => E, P => O	36.	L => E, J => N, Q => M
9.	S => A, E => L, Y => R	37.	E => S, V => M, X => H
10.	B => R, J => T, U => A	38.	Y => G, C => O, W => A
11.	T => H, V => W, R => E	39.	M => D, P => E, V => A
12.	C => A, J => U, G => R	40.	V => I, M => S, R => L
13.	G => E, W => O, U => P	41.	
14.	L => T, O => R, Z => A	42.	
15.	B => F, L => A, U => C	43.	
16.	X => E, C => L, N => I	44.	
17.	J => O, S => R, G => C	45.	
18.	R => E, V => F, Z => W	46.	
19.	F => I, B => O, K => H	47.	
20.	G => N, Y => O, R => C	48.	
21.	S => H, H => A, M => L	49.	
22.	W => E, Y => D, P => N	50.	
23.	R => A, E => R, V => E	51.	
24.	L => R, N => E, M => C	52.	
25.	T => R, Y => N, U => V	53.	
26.	N => T, S => A, H => E	54.	
27.	N => G, A => I, H => M	55.	
28.	U => O, L => T, Q => N	56.	

CRYPTOGRAM - ANSWERS

1.	Ants do not sleep
	--
2.	Uranus has 27 moons
	--
3.	Earthworms have 5 hearts
	--
4.	Polar bears are left-handed
	--
5.	There is cyanide in apple pips
	--
6.	A human eyeball weighs an ounce
	--
7.	Adolf Hitler loved chocolate cake
	--
8.	Mr. Rogers is an ordained minister
	--
9.	Oak trees can live 200 or more years
	--
10.	Persia changed its name to Iran in 1935
	--
11.	The most popular grown bulbs are tulips
	--
12.	A group of larks is called an exaltation
	--
13.	On average, pigs live for about 15 years
	--
14.	The Hawaiian alphabet only has 12 letters
	--
15.	A cow has four compartments in its stomach
	--
16.	Every day 2,700 people die of heart disease
	--

CRYPTOGRAM - ANSWERS

17.	Over 90% of poison exposures occur in homes	--
18.	There are over 1,800 known species of fleas	--
19.	Urophobia is the fear of urine or urinating	--
20.	In Las Vegas, casinos do not have any clocks	--
21.	The most popular brand of raisins is Sunmaid	--
22.	Benjamin Franklin invented the rocking chair.	--
23.	Every 30 seconds a house fire doubles in size	--
24.	One out of four American households own a cat	--
25.	There are over 500 different types of bananas	--
26.	There are 293 ways to make change for a dollar	--
27.	Twit is the name given for a pregnant goldfish	--
28.	Twelve men have landed on and explored the moon	--
29.	Between 12%-15% of the population is left-handed	--
30.	Los Angeles is the most polluted city in the USA	--
31.	There are more than 640 muscles in the human body	--
32.	500,000 kids in the US live in same sex households	--

CRYPTOGRAM - ANSWERS ( )

33.	Copper is the second most used metal in the world.
	--
34.	Only 4% of babies are born on their actual due date
	--
35.	75-90% of primary physician visits are due to stress
	--
36.	In an year, an average person makes 1,140 phone calls
	--
37.	dous:tremendous, horrendous, stupendous, and hazardous
	--
38.	On average people fear spiders more than they do death
	--
39.	The space between your eyebrows is called the Glabella
	--
40.	There are approximately 75,000,000 horses in the world
	--
41.	
	--
42.	
	--
43.	
	--
44.	
	--
45.	
	--
46.	
	--
47.	
	--
48.	
	--

INTERESTING FACTS

1.

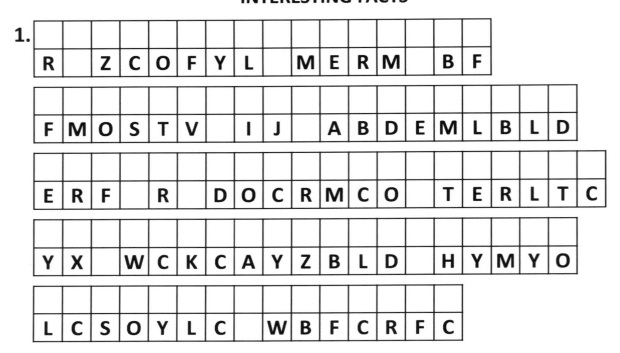

| | R | | Z | C | O | F | Y | L | | M | E | R | M | | B | F |

| | F | M | O | S | T | V | | I | J | | A | B | D | E | M | L | B | L | D |

| | E | R | F | | R | | D | O | C | R | M | C | O | | T | E | R | L | T | C |

| | Y | X | | W | C | K | C | A | Y | Z | B | L | D | | H | Y | M | Y | O |

| | L | C | S | O | Y | L | C | | W | B | F | C | R | F | C |

2.

| | L | Q | C | | A | I | C | S | L | | M | T | Y | C | L | | T | Z |

| | 1 | 8 | 4 | 3 | | Q | S | E | | S | | L | S | V | K | | L | Q | S | L |

| | D | S | W | | T | X | C | I | | 3 | 0 | 0 |

| | F | V | K | T | Y | C | L | I | C | W | | K | T | U | A | . |

INTERESTING FACTS

3.

J	F	M		V	L	R	A	M	Z	C	R	J	I		G	T

E	X	I	W	G	V	J	F		O	K	C		J	F	M

T	R	Z	C	J		V	L	R	A	M	Z	C	R	J	I		J	G

G	T	T	M	Z		K		N	M	P	Z	M	M		R	L

C	V	Z	T	R	L	P

INTERESTING FACTS

4.

R	E	B		O	S	S		D	I		R	E	B		X	G	Y	Y	B	H

D	N		O	D	J	H		F	B	Q	G	K		K	R	L	I	C	K

N	D	H		O	D	K	E	G	C	L		S	D	W	O	D

S	L	A	J	K	E	G	A	G	A	L	G	K	E	L	,		R	E	B

T	D	H	F	C	K		F	L	H	W	B	K	R		X	G	Y	Y	B	H

U	L	I	J	N	L	V	R	J	H	B	H

INTERESTING FACTS

5.

| K | S | P | | V | X | D | N | K | | K | P | A | P | H | X | N | X | M | U |

| N | S | M | B | | K | M | | N | S | M | B | | J | U | G |

| Q | M | D | K | X | M | U | | M | V | | J | | K | M | X | A | P | K | | B | J | N |

| M | U | | " | A | P | J | H | P | | X | K | | K | M |

| Y | P | J | H | P | D | . | " | | J | V | K | P | D | | V | X | F | S | K | X | U | F |

| V | M | D | | K | P | U | | B | P | P | E | N | | K | M | | N | S | M | B |

| K | S | P | | K | M | X | A | P | K | , | | I | Y | N | | B | M | C | A | Z |

| M | U | A | G | | J | A | A | M | B | | K | S | P |

| Q | D | M | Z | C | I | P | D | N | | K | M | | N | S | M | B | | K | S | P |

| K | M | X | A | P | K | | K | J | U | E | , | | J | U | Z | | U | M | K |

| K | S | P | | B | S | M | A | P | | K | M | X | A | P | K |

--

INTERESTING FACTS

6.

| A | F | T | | K | G | V | E | T | | U | Z | | G |

| X | T | W | N | T | M | G | P | | P | U | V | E | - | O | U | C | | N | J |

| 2 | 2 | 0 | | I | G | K | W | J |

7.

| E | F | C | | P | S | O | U | T | P | , | | A | F | U | I | F | | U | L |

| W | T | E | U | H | C | | E | S | | T | Z | T | L | P | T | , | | U | L |

| E | F | C | | Z | T | X | Q | C | L | E | | D | C | T | X | | T | W | O |

| I | T | W | | B | C | T | L | J | X | C | | J | Y | | E | S |

| C | U | Q | F | E | | R | C | C | E | | T | W | O | | A | C | U | Q | F |

| T | L | | B | J | I | F | | T | L | | 1 | , | 7 | 0 | 0 |

| Y | S | J | W | O | L |

INTERESTING FACTS

8.

NFC ZUIEB JVFEIJUME

LMYCLENFY UVNWMU

RCRML IZCYM,

CKMLNJCFZ CLM ZCNE UV

IZM UOM KVZU UVNWMU

RCRML RML ULNR UV UOM

GCUOLVVK, QONJO QCZ

ZMAMF ZOMMUZ VH

UVNWMU RCRML RML ULNR --

INTERESTING FACTS

9.

W	H		H	B	V	T	K

B	I	I	O	E	Z	W	C	B	H	T	G	J		1	2

J	T	B	O	K		L	E	O		S	X	I	W	H	T	O		H	E

E	O	A	W	H		H	P	T		K	X	U

--

10.

H	D	L		Q	S	H	J	D		Q	S	I

K	G	P	L	G	H	L	F		W	X		R	L	H	L	Z

D	L	G	O	L	K	G		U	M		G	V	Z	L	E	W	L	Z	T

K	G		1	5	1	0	.

--

INTERESTING FACTS

11.

F S C D C   R Q T   Q   A L T F

L B B G K C   L Z   F S C   D M T T G Q Z

T A Q K C   T F Q F G L Z   I G D .

P G T G F G Z E   K L T I L Z Q M F T

R L M N J   M T C   M Z G W M C

A L T F Q N   " I Q D H C D T "   F L

T F Q I A   C Z P C N L A C T   Q Z J

L F S C D   G F C I T   Q T   S Q P G Z E

B N L R Z   Q Y L Q D J   F S C   I G D

T A Q K C   T F Q F G L Z

INTERESTING FACTS

12.

| V | U | J | | K | N | R | U | V | | B | D | O | R | | T | A | | M |

| U | D | S | M | O | | N | W | | B | M | K | R | J | K | | V | U | M | O |

| V | U | J | | B | J | A | V | | T | O | J | . | | V | U | N | W | | N | W |

| I | J | L | M | D | W | J | | T | A | | V | U | J | | W | F | M | L | J |

| M | O | Q | | F | B | M | L | J | S | J | O | V | | T | A | | V | U | J |

| U | J | M | K | V |

13.

| Q | A | H | | M | B | P | Z | Q | | U | Y | H | P | B | I | U | T |

| I | H | V | H | N | P | U | Q | B | F | T | | F | M | | Z | Q | . |

| J | U | Q | P | B | I | S | Z | | E | U | G | | O | U | Z | | U | Q |

| N | F | Z | Q | F | T | | B | T | | 1 | 7 | 3 | 7 |

INTERESTING FACTS

14.

| Y | P | O | | I | V | H | R | Y | | N | K | T | V | X | | V | N |

| Y | P | O | | E | H | B | I | O | N | Y | | H | R | V | A | H | E |

| Z | V | Y | P | Q | T | Y | | H | | J | H | W | C | J | Q | R | O |

--

15.

| Y | A | X | | Y | Z | | U | R | X | | F | Y | A | M | | T | W | N | G | Q |

| W | A | | U | R | X | | B | E | J | X | Q | | F | Y | A | M |

| J | Y | K | W | X | , | | " | Z | Y | N | | I | Y | V | N | | X | I | X | Q |

| Y | A | G | I | , | " | | V | Q | X | M | | U | Y | | F | X | E |

| J | E | A |

--

INTERESTING FACTS

16.

Q	S	G		C	M	H	F	Q		F	N	K	P	T	B

F	B	F	Q	G	A		M	W		T	A	G	H	M	V	T		P	T	F

K	N	M	D	Q		M	W		K	Z	F	Q	Z	W	,

A	T	F	F	T	V	S	N	F	G	Q	Q	F		M	W		1	8	9	7

--

17.

P	I	F	Z		U	P	E	Z	W	I	H	K	P		C	U	Y	Z

I	E	C	Z	W		U	P	E	Z	W	I	H	K	P

I	W	A	H	E	H	R	M		E	C	Z	F

--

INTERESTING FACTS

18.

R	C	Z		N	Y	R	Z	J	V	Y	U		R	W		X	M	V	U	K

R	C	Z		R	Y	S		N	Y	C	Y	U		T	Y	A

X	J	W	M	Q	C	R		V	E		P	J	W	N

H	Y	J	V	W	M	A		O	Y	J	R	A		W	P

V	E	K	V	Y		X	I		Y		P	U	Z	Z	R		W	P

1	0	0	0		Z	U	Z	O	C	Y	E	R	A

--

INTERESTING FACTS

19.

O	G	G		C	A	Y		Q	T	Z	V	L	B		Q	T	Y	U	Y

N	X	M	S	C		V	C	X	T	Y	V

P	X	G	G	Y	P	C	Z	W	Y	G	B		P	X	S	G	N

U	O	Q	Y		O		N	X	S	H	A	M	S	C		V	C	O	P	Q

O	V		A	Z	H	A		O	V		C	A	Y		Y	U	L	Z	T	Y

V	C	O	C	Y		I	S	Z	G	N	Z	M	H		Z	M

X	M	G	B		2		U	Z	M	S	C	Y	V

--

20.

L	Z	C		"	S	C	F	P	U	Y	J		Z	Y	L

R	Y	J	U	C	"		P	Q		L	Z	C

H	D	D	P	U	P	Y	W		R	Y	J	U	C		H	D

S	C	F	P	U	H

--

INTERESTING FACTS

21. | V | O | Z | F | Z | | W | Q | | U | D | | V | W | S | S | W | U | X |

| W | U | | W | J | Z | R | K | U | E |

--

22. | W | H | | Y | G | H | E | C | A | Y | V | , | | 5 | 0 |

| D | G | Z | A | G | H | E | | U | B | | E | T | G | | D | G | U | D | N | G |

| O | T | U | | J | G | E | | R | Q | Z | Z | W | G | P | | B | U | Z |

| E | T | G | | B | W | Z | I | E | | E | W | R | G | | Q | Z | G |

| E | G | G | H | Q | J | G | Z | I |

--

INTERESTING FACTS

23.

J C Q B U   Q I B   K B J Q I   D C

Q I B   M B X G A W ,   J O S B U Q

B G X W Q B G X ,   I G W   S U J G X

N J W   U B H D P B K   S L   J

T J Q I D O D M G W Q   J X K   T A Q

G X   J   V J U   C D U   C A Q A U B

W Q A K L

INTERESTING FACTS

24.

| Y | H | X | J | | D | I | H | X | H | V | | Y | H | O | D |

| Z | W | J | L | O | | Z | J | O | O | L | Z | I | O | U | | M | U |

| O | N | M | M | L | E | P | | Z | W | J | L | O | | S | W | J | V | Z | V |

| I | E | | H | | Z | O | J | J | . | | Y | H | X | J |

| D | I | H | X | H | V | | W | H | T | J | | H | | F | H | O | D |

| V | S | J | E | Z | | P | X | H | E | F | | L | E | | Z | W | J |

| Y | L | F | F | X | J | | I | A | | Z | W | J | L | O | | S | W | J | V | Z |

--

INTERESTING FACTS

25.

| S | V | L | E | L | | P | J | | T | | A | T | E | M | L |

| W | E | T | J | J | | J | S | T | S | Z | L | | K | G |

| Q | P | I | I | P | L | - | S | V | L | - | F | K | K | V | | P | I |

| A | P | X | T | , | | F | L | E | Z |

26.

| P | S | L | | Q | P | L | L | O | L | Q | P | | Q | P | W | L | L | P |

| F | M | | P | S | L | | C | Z | W | U | A | | F | Q |

| I | E | U | A | C | F | M | | Q | P | W | L | L | P |

| U | Z | V | E | P | L | A | | F | M | | A | B | M | L | A | F | M | , |

| M | L | C | | H | L | E | U | E | M | A | . | | F | P | | S | E | Q |

| E | M | | F | M | V | U | F | M | L | | Z | X | | 3 | 8 | % |

INTERESTING FACTS

27.

| G | D | | W | N | Z | W | D | G | W | , | | D | L | Y | Y | G | D | U |

| J | L | F | M | | I | P | W | Y | | H | P | W | D | R | | " | D | L | " |

| W | D | Y | | R | I | W | T | G | D | U | | J | L | F | M | | I | P | W | Y |

| H | P | W | D | R | | " | J | P | R | . | " |

--

28.

| B | V | M | S | P | | Y | P | Q | G | X | Y | | N | S | A |

| " | B | O | V | B | A | D | T | B | G | P | Q | C |

| V | A | S | P | S | P | F | V | B | | M | D | P | F | | S | N |

| P | S | T | S | A | A | S | K | . | " |

--

INTERESTING FACTS

29.

Q	U	K	F		Y	P	B	W	X		Q	B	U	V	E	Q	O	Z

U	W		Z	B	V	U	W

INTERESTING FACTS

30.

L E K S T I I   H U E Q T R R T

J M T U M M T S   F L I   K Q T

M U S I K   E Q W U E T   K W   J R L B

E R L S U E T   I K L S R U P D   U P

K Q T   H W Y U T   " I U R T P E T   W M

K Q T   R L H Z I . "   I Q T

K N S P T V   V W F P   K Q T   S W R T

Z T E L N I T   I Q T   M W N P V   U K

K W W   I E L S B

--

INTERESTING FACTS

31.

O I N U R | Q N P E | Q S N T P V

" 9 9 - 4 4 / 1 0 0 % | E Y U K " | Z P Q

X S K I K U S R | O V I K V B K F | G R

J P U S K R | E U N X B N U | Z J N

Z O B J | B J K | J K S E | N M

X J K A O Q B Q | F K B K U A O V K F

B J P B | O I N U R | Q N P E | Z P Q

N V S R | 5 6 / 1 0 0 | E Y U K .

E U N X B N U | Q O A E S R

Q Y G B U P X B K F | 5 6 | M U N A | 1 0 0

P V F | X P A K | Y E | Z O B J

" 9 9 - 4 4 / 1 0 0 % | E Y U K "

--

INTERESTING FACTS

32.

N	L	I		Z	F	I	C	Z	X	I		I	Z	C		X	C	R	V	U

0	.	0	1		Y	G	M	L	I	U		Y	G		H	I	G	X	N	L

I	F	I	C	J		J	I	Z	C

--

33.

V	Q	C		U	V	J	V	C		W	M		J	O	J	U	L	J

Q	J	U		J	O	A	W	U	V		V	N	H	P	C		J	U

A	J	D	G		P	J	Z	H	K	W	X		J	U

S	C	W	S	O	C

--

INTERESTING FACTS

34.

| S | J | X | W | X | | L | B | | Z | | 1 | | L | M | | 4 |

| F | J | Z | M | F | X | | S | J | Z | S | | M | X | G | | A | E | W | V |

| G | L | K | K | | J | Z | U | X | | Z | | G | J | L | S | X |

| F | J | W | L | B | S | N | Z | B | . |

--

35.

| H | E | I | | P | R | | W | L | C | | R | C | J | E | Y | X |

| B | E | R | W | | R | K | E | G | C | Y | | C | Y | Q | S | P | R | L |

| O | E | U | X |

--

INTERESTING FACTS

36.

U E Z   X K I B Z A Z V U   G C   I

T G V L Z O ' Y   Z O Z Y   M V

M U ' Y '   E Z I T Y   Z V I P K Z Y

M U   U G   Y Z Z   I K K   C G N D

C Z Z U   I U   I K K   U M A Z Y !

--

37.

F J U   I Q Q C   I B   P U Y G L

P I D G   S W M U   W D U   W P I O F

5 1 . 2 % ,   W E E I D Q Y G L   F I

E U G C O C .

--

INTERESTING FACTS

38.

V	N	J	F	K		A	D	L	K	I		K	F	U	E	D	X

Z	V	L	O	F		U	H		J	Y	O	W		U	H		J	F	K	.

39.

S	W	O		I	P	A	O		Z	O	I	L	Q		Z	P	V

A	P	L	O		R	M		C	X	H		S	W	O		T	X	X	G

M	O	S	O	H		M	P	I	.		S	W	O	H	O		Z	P	V

I	O	N	O	H		P		H	O	B	X	H	L	O	L

Z	O	I	L	Q		T	O	C	X	H	O		U	S	.

INTERESTING FACTS - HINTS

1.	B => I, R => A, D => G	**21.**	W => I, X => G, U => N
2.	Z => F, L => T, M => C	**22.**	V => Y, W => I, Z => R
3.	T => F, J => T, M => E	**23.**	H => M, D => O, J => A
4.	G => I, R => T, K => S	**24.**	J => E, E => N, F => D
5.	U => N, Y => B, K => T	**25.**	P => I, S => T, J => S
6.	V => N, G => A, E => G	**26.**	P => T, M => N, C => W
7.	Q => G, S => O, L => S	**27.**	G => I, F => U, R => S
8.	U => T, L => R, C => A	**28.**	V => P, A => R, C => L
9.	E => O, H => T, T => E	**29.**	Q => F, E => H, P => E
10.	L => E, Z => R, F => D	**30.**	T => E, S => R, U => I
11.	P => V, C => E, K => C	**31.**	N => O, S => L, U => R
12.	V => T, U => H, M => A	**32.**	I => E, U => S, C => R
13.	M => F, Q => T, O => W	**33.**	C => E, O => L, N => W
14.	K => Q, P => H, H => A	**34.**	M => N, G => W, F => C
15.	Q => S, N => R, Y => O	**35.**	L => H, C => E, W => T
16.	K => B, B => Y, F => S	**36.**	U => T, O => Y, V => N
17.	K => D, W => R, F => M	**37.**	I => O, W => A, Q => D
18.	W => O, K => D, M => U	**38.**	O => C, H => S, U => A
19.	X => O, Y => E, C => T	**39.**	M => P, O => E, H => R
20.	Z => H, L => T, C => E		

INTERESTING FACTS - ANSWERS (1/3)

1.	A person that is struck by lightning has a greater chance of developing motor neurone disease
2.	The Great Comet of 1843 had a tail that was over 300 kilometres long.
3.	The University of Plymouth was the first university to offer a degree in surfing
4.	The YKK on the zipper of your Levis stands for Yoshida Kogyo Kabushibibaisha, the worlds largest zipper manufacturer
5.	The first television show to show any portion of a toilet was on "Leave it to Beaver." After fighting for ten weeks to show the toilet, CBS would only allow the producers to show the toilet tank, and not the whole toilet
6.	The range of a medieval long-bow is 220 yards
7.	The Kodiak, which is native to Alaska, is the largest bear and can measure up to eight feet and weigh as much as 1,700 pounds
8.	In a study conducted regarding toilet paper usage, Americans are said to use the most toilet paper per trip to the bathroom, which was seven sheets of toilet paper per trip
9.	It takes approximately 12 years for Jupiter to orbit the sun
10.	The watch was invented by Peter Henlein of Nuremberg in 1510.
11.	There was a post office on the Russian space station Mir. Visiting cosmonauts would use unique postal "markers" to stamp envelopes and other items as having flown aboard the Mir space station
12.	The right lung of a human is larger than the left one. This is because of the space and placement of the heart
13.	The first American celebration of St. Patricks Day was at Boston in 1737
14.	The giant squid is the largest animal without a backbone

15.	One of the Bond girls in the James Bond movie, "For Your Eyes Only," used to be a man --
16.	The first subway system in America was built in Boston, Massachusetts in 1897 --
17.	Some asteroids have other asteroids orbiting them --
18.	The material to build the Taj Mahal was brought in from various parts of India by a fleet of 1000 elephants --
19.	All the Krispy Kreme donut stores collectively could make a doughnut stack as high as the Empire State Building in only 2 minutes --
20.	The "Mexican Hat Dance" is the official dance of Mexico --
21.	There is no tipping in Iceland --
22.	In Kentucky, 50 percent of the people who get married for the first time are teenagers --
23.	After the death of the genius, Albert Einstein, his brain was removed by a pathologist and put in a jar for future study --
24.	Male koalas mark their territory by rubbing their chests on a tree. Male koalas have a dark scent gland in the middle of their chest --
25.	There is a large brass statue of Winnie-the-Pooh in Lima, Peru --
26.	The steepest street in the world is Baldwin Street located in Dunedin, New Zealand. It has an incline of 38% --
27.	In Albania, nodding your head means "no" and shaking your head means "yes." --
28.	EPCOT stands for "Experimental Prototype City Of Tomorrow." --

INTERESTING FACTS - ANSWERS (3/3)

29.	Fido means faithful in Latin
30.	Actress Michelle Pfeiffer was the first choice to play Clarice Starling in the movie "Silence of the Lambs." She turned down the role because she found it too scary
31.	Ivory soap slogan "99-44/100% Pure" was cleverly invented by Harley Proctor who with the help of chemists determined that Ivory soap was only 56/100 pure. Proctor simply subtracted 56 from 100 and came up with "99-44/100% Pure"
32.	The average ear grows 0.01 inches in length every year
33.	The state of Alaska has almost twice as many caribou as people
34.	There is a 1 in 4 chance that New York will have a white Christmas.
35.	You is the second most spoken English word
36.	The placement of a donkey's eyes in it's' heads enables it to see all four feet at all times!
37.	The odds of being born male are about 51.2%, according to census.
38.	Women blink nearly twice as much as men.
39.	The name Wendy was made up for the book Peter Pan. There was never a recorded Wendy before it.

CRYPTOGRAM

1.

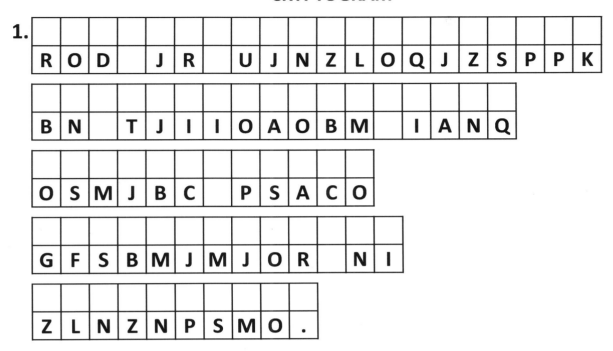

| R | O | D | | J | R | | U | J | N | Z | L | O | Q | J | Z | S | P | P | K |

| B | N | | T | J | I | I | O | A | O | B | M | | I | A | N | Q |

| O | S | M | J | B | C | | P | S | A | C | O |

| G | F | S | B | M | J | M | J | O | R | | N | I |

| Z | L | N | Z | N | P | S | M | O | . |

--

2.

| L | | D | H | L | F | S | D | P |

| (| A | R | A | - | W | R | F | R | N | V | F | X | A | E |) | | D | B | G | L | A |

| H | P | H | | W | L | A | | E | X | J | S | X | A | Q | B | X | J | D |

| V | H | S | T | H | H | A | | 5 | 0 | 0 | | J | D | L | E | H | J | | R | I |

| Q | N | L | P | . |

--

CRYPTOGRAM

3.

A Z   1 9 8 4 ,   H   Q H Z H G A H Z

X H Y D C Y   K C B H Z   Y C Z S A Z B

H G N C Y S A R A Z B   R E H Q C   L Z

T A R   Q L J R .

4.

J   I U B Y C W X I C G   U C I Y   B G

L J E X   C D   W C L T J W Z X E

U J B I .

5.

V H   V P   V R B E P P V T S O   H E

S V F A   W E I K   O S T E D .

CRYPTOGRAM

6.

T J | K Z C | A N N D | H | W Z I U J T L Y

T P | G Y N | U H X A | X Z Z O , | T G

M T I I | N R N P G C H I I K | G C X P

M Y T G N .

7.

P Q Z L | F U S | G L Z Z J Z , | O N N

F U S C | K U I Y N F | X S L D R Y U L G

G R U H | Z W Z L | F U S C | Q Z O C R .

8.

O | F O R | B O V | 3 2 | K Q V F J G V

W E | G O F B | G O H .

CRYPTOGRAM

9.

S VRUSC OLSCP PWEG EP

OUSZW. EBP NDC EP

SZBDSUUH ZULSC, ODB

UEWL PGRK EB SVVLSCP

KYEBL.

10.

RIOEPTYD KTLS QXWAYT

OD AIY WHTODKAOUY.

CRYPTOGRAM

11. L C Q T . Y . U Z L K V Z L C L C Q

C Z J C Q Y L P S L Q B I

N Z J C L D Z D J Y L P Z H Q Y F Q P

U S F Z L S Z Y U N Q S P V S L Q P ,

I N B P Z G S .

--

12. E K T L E Z T K T M M T G N D Z Z

X D T D K D G Y U T R D P G U

C T E G E P X I E P P U G K D P X E

L E G T .

--

CRYPTOGRAM

13.

| D | P | R | ' | L | Z | | T | P | L | U | | B | W | G | E | | 3 | 0 | 0 |

| T | P | U | Z | N | , | | T | R | G | | T | D | | G | E | Z | | G | W | O | Z |

| D | P | R | | T | Z | S | P | O | Z | | F | U | | F | K | R | I | G | , |

| D | P | R | | P | U | I | D | | E | F | Y | Z | | 2 | 0 | 6 | . |

--

14.

| X | Q | J | V | G | | I | C | G | Q | E | X | M | | B | G | | Y | P | Q |

| Y | Q | R | P | J | B | R | M | H | | Y | Q | E | O | | S | I | E | | Y | P | Q |

| " | O | B | G | G | B | I | J | M | E | U | | T | I | G | B | Y | B | I | J | . | " |

--

CRYPTOGRAM

15.

| N | W | Z | | R | Q | N | I | S | R | N | Z | Y | | N | Z | T | T | Z | A |

| S | R | X | W | V | B | Z | | (| R | N | S |) | | J | R | P |

| V | B | N | A | I | Y | Q | X | Z | Y | | V | B | | Z | B | M | T | R | B | Y |

| V | B | | 1 | 9 | 6 | 5 | . |

--

16.

| N | W | Q | Q | C | B | | K | C | W | Z | Z | S | | W | B | M |

| X | Y | R | Q | O | C | S | | F | L | O | W | R | B | C | | W | Q | C |

| K | Q | E | Z | Y | C | Q | | W | B | M | | X | R | X | Z | C | Q | . |

--

CRYPTOGRAM

17.

BCKDVBCE PULC ZCCT

YTWNT BW CUB AWWX

BNVOC UE AUEB NPCT

PCULJ DCBUM DHEVO VE

GMUJVTR .

--

18.

WN MFQKQHZ NCPM DUZK

KDCJDN FHL DNQHZ CH

NCPM WFTB , NCP TFHHCR

KQHB QHRC YPQTBKFHL .

--

CRYPTOGRAM

19.

| F | G | X | | P | T | Y | R | F | | I | U | K | I | | F | G | X |

| V | W | W | T | X | R | | S | Y | U | H | H | X | S | | U | A |

| I | X | Y | W | T | A | | T | A | | N | N | T | T | | L | T | W | W | X | S |

| F | G | X | | U | A | W | D | | X | W | X | H | G | V | A | F | | T | A |

| F | G | X | | I | X | Y | W | T | A | | J | U | U | . |

20.

| " | S | F | J | E | H | | S | F | J | P | " | | X | U | K | | U |

| $ | 2 | 5 | , | 0 | 0 | 0 | , | 0 | 0 | 0 | | K | U | Y | U | O | P | , |

| Q | X | L | Y | H | | K | F | R | O | H | T | H | | V | B | F | O | N |

| S | F | K | N | L | V | H | | E | L | I | K | G | H | O | E | | X | U | K | | U |

| $ | 1 | 9 | 0 | , | 1 | 0 | 0 | | K | U | Y | U | O | P | . |

CRYPTOGRAM

21.

AWP GDPXGOP QPXZST

JGFFZ GZFPPQ RT ZPDPT

IRTBAPZ.

--

22.

NFD "HXQQ IDY'C FKQQ

ZP PKID " TC QZJKNDH TY

JKLLZQQ, VTCJZYCTY.

--

CRYPTOGRAM

23.

A B N B U T V E Y Z V T B J N Y B A L X R Y B

S V B H U   Q V B N   X Q   Q N Y Z B P   J L V

1 3 J L ,   G L Y F L   X F F M N U   X H V

J X   J L N V V   J Y S V U   B   P V B N .

--

24.

D F K Z K   W Z K   W C Y L D   W

P B T T B Y Q   W Q D O   M K Z

M K Z O Y Q .   W Q D O   W Z K   G K Z I

O Y R B W T   W Q B P W T O   W Q J

U B T T   T B G K   B Q   R Y T Y Q B K O

D F W D   R W Q   R Y Q D W B Q

W T P Y O D   5 0 0 , 0 0 0   W Q D O .

--

CRYPTOGRAM

25.

JOPJXO OWV SIMOBVM – BWXXOC

"OIVPLPJTWKU" (JOPJXO OWVSIK

RHKM) – SV TWM ROOI JNWBVSBOC

ZPN BOIVHNSOM VTNPHKTPHV

WZNSBW, WHMVNWXSW, WMSW, VTO

LSCCXO OWMV, WIC IPNVT,

BOIVNWX WIC MPHVT WLONSBW.

YTU? ROBWHMO LWIU RHKM WNO

RPVT JNPVOSI-NSBT WIC KPPC

MPHNBOM PZ ASVWLSIM, LSIONWXM

WIC ZWVM.

--

CRYPTOGRAM

26.

YPD ZIDVZSD APDWB-WKBD

EB Z WZYDL MEQOEG KA

ZUEHY YRE TDZVA.

27.

HGOTHN-IOAJ CJTRJDH FI

HGJ CJFCPJ SGF YZJ

CJTZFDQP QKZ IFT

KQHODU QTJ QPTJQKN

XQTTOJK.

CRYPTOGRAM

28.

YZ BLYOHBQ CS FMH KCQF

LBG FMHBH YQ L ELKK

DLBWYZJ CS B2-O2 LZO

D-3X0 AHMYZO FMH LBG

--

29.

IQB FWBEFUB TFDFLBJB

QSVJBQSRX GFIMQBJ PSEB

IQFL 10 QSVEJ SO

IBRBWHJHSL F XFC.

--

CRYPTOGRAM

30.

| Q | S | H | K | | Y | S | H | | Y | R | Y | L | K | R | G | | E | L | K | N | , |

| 2 | 2 | 2 | 8 | | W | H | U | W | T | H | | Q | H | F | H | | U | K |

| R | Y | . | | U | K | T | X | | 7 | 0 | 6 | | E | V | F | D | R | D | H | M | . |

--

31.

| D | Z | V | X | M | F | D | S | V | | F | I | B | J |

| D | X | A | Z | V | X | M | F | H | | P | B | X | Z | V |

| Z | V | I | L | L | I | N | | M | F | | V | T | I | | A | M | M | F |

| R | B | V | T | | T | B | Z | | J | I | P | V | | P | M | M | V | . |

--

CRYPTOGRAM

32.

YIK TOJIKROW TMW OL

SWPBW ML TD. EOE. IOL

WMTK BML EOEKWGNT OW

YIK JPTXMWZ'L VODLY

MGLOW 1896.

--

33.

ZRNJ YNXND GBOU "PNBQ

QN TC, GFAZZJ" AY GZBD

ZDNV.

--

CRYPTOGRAM

34.

V L   L C P D U   B N K D   E C S N K V D U

L N   D C L   C   O V D E D   N W

E D S D K T   L A C X   L A D   E D S D K T

A C U   V X   V L   L N   I D Y V X

G V L A .

--

35.

Q G L L G U   Q G N Y B   A R U G L   O D

U G K   G U   K F R   I O D K   G S   K G M

1 0   A R U G L D   H R K   O K   O D

D K O I I   4 0   K O L R D   L G Y R

K G W O Q   K F B U   Q H B U O C R .

--

CRYPTOGRAM

36.

K	N	X		N	H	T	D	Q		C	X	X	K

B	X	J	Y	B	A	J	X		N	D	I	C		D		B	A	Q	K

E	C		C	I	H	A	V		D		V	D	U

--

37.

R	J	G	S	J	H		Z	V	A	H	V	N	'	H

V	J	Q	W	D	N	V	C	O		S	H		F	W	N	Q	O

V	I	W	J	Q		$	2	,	0	0	0	,	0	0	0	.

--

CRYPTOGRAM

38.

X K A   F D G Z V G   L K E Z B

P V K G D J   B F D   M V E J

" A Y H H D B " .   K B   K Z   Q

P V A C K G Q B K V G   V L

" A Q E K V G D B B D "   Q G J

" H Y H H D B . "

39.

K Y   K J Q G K W Q   M Z T K Y   C I O Q O

K X I Z D   2 0 0   M Q K V   M K S G O

B Q G   V K L .

CRYPTOGRAM

40.

B	V	U		F	B	I	B	U		K	R		R	X	K	Q	N	C	I

N	F		J	N	Z	Z	U	Q		B	V	I	D

U	D	Z	X	I	D	C	.

CRYPTOGRAM - HINTS

1.	Z => C, S => A, O => E	21.	F => L, R => I, A => T
2.	A => N, L => A, V => B	22.	Q => L, P => F, Z => O
3.	K => B, Z => N, H => A	23.	B => A, Q => F, H => N
4.	G => S, I => R, D => F	24.	K => E, W => A, C => B
5.	E => O, W => Y, T => B	25.	I => N, P => O, U => Y
6.	G => T, P => N, X => R	26.	B => F, Z => A, V => R
7.	Z => E, Q => H, G => S	27.	F => O, X => M, N => Y
8.	F => C, O => A, G => E	28.	Q => S, G => K, M => H
9.	U => L, D => U, H => Y	29.	B => E, F => A, Q => H
10.	I => H, O => I, D => N	30.	H => E, K => N, U => O
11.	L => T, Z => I, Q => E	31.	R => W, Z => S, V => T
12.	D => I, N => W, L => M	32.	N => U, R => L, L => S
13.	G => T, P => O, R => U	33.	Z => T, N => E, A => O
14.	T => P, B => I, E => R	34.	A => H, D => E, N => O
15.	N => T, T => L, Y => D	35.	B => A, K => T, O => I
16.	Y => H, Q => R, X => S	36.	T => M, X => E, K => T
17.	E => S, B => T, W => O	37.	H => S, V => A, G => L
18.	P => U, B => K, L => D	38.	Z => S, K => I, F => H
19.	J => Z, W => L, G => H	39.	Q => E, G => R, M => H
20.	F => U, S => J, B => O	40.	I => A, Z => G, R => F

CRYPTOGRAM - ANSWERS (1/3)

1.	Sex is biochemically no different from eating large quantities of chocolate.
	--
2.	A healthy (non-colorblind) human eye can distinguish between 500 shades of gray.
	--
3.	In 1984, a Canadian farmer began renting advertising space on his cows.
	--
4.	A rhinoceros horn is made of compacted hair.
	--
5.	It is impossible to lick your elbow.
	--
6.	If you keep a goldfish in the dark room, it will eventually turn white.
	--
7.	When you sneeze, all your bodily functions stop even your heart.
	--
8.	A cat has 32 muscles in each ear.
	--
9.	A polar bears skin is black. Its fur is actually clear, but like snow it appears white.
	--
10.	Children grow faster in the springtime.
	--
11.	The U.S. city with the highest rate of lightning strikes per capita is Clearwater, Florida.
	--
12.	A female ferret will die if it goes into heat and cannot find a mate.
	--
13.	You're born with 300 bones, but by the time you become an adult, you only have 206.
	--
14.	Venus observa is the technical term for the "missionary position."
	--
15.	The Automated Teller Machine (ATM) was introduced in England in 1965.
	--
16.	Warren Beatty and Shirley McLaine are brother and sister.
	--
17.	Termites have been known to eat food twice as fast when heavy metal music is playing.
	--

18.	By raising your legs slowly and lying on your back, you cannot sink into quicksand.
19.	The first bomb the Allies dropped on Berlin in WWII killed the only elephant in the Berlin Zoo.
20.	"Judge Judy" has a $25,000,000 salary, while Supreme Court Justice Ginsberg has a $190,100 salary.
21.	The average person falls asleep in seven minutes.
22.	The "Dull Men's Hall of Fame" is located in Carroll, Wisconsin.
23.	Paraskevidekatriaphobia means fear of Friday the 13th, which occurs one to three times a year.
24.	There are about a million ants per person. Ants are very social animals and will live in colonies that can contain almost 500,000 ants.
25.	People eat insects – called "Entomophagy"(people eating bugs) – it has been practiced for centuries throughout Africa, Australia, Asia, the Middle East, and North, Central and South America. Why? Because many bugs are both protein-rich and good sources of vitamins, minerals and fats.
26.	The average shelf-life of a latex condom is about two years.
27.	Thirty-five percent of the people who use personal ads for dating are already married.
28.	In Raiders of the Lost Ark there is a wall carving of R2-D2 and C-3P0 behind the ark
29.	The average Japanese household watches more than 10 hours of television a day.
30.	When the Titanic sank, 2228 people were on it. Only 706 survived.
31.	Astronaut Neil Armstrong first stepped on the moon with his left foot.
32.	The Michelin man is known as Mr. Bib. His name was Bibendum in the company's first ads in 1896.
33.	They NEVER said "Beam me up, Scotty" on Star Trek.

CRYPTOGRAM - ANSWERS (3/3)

34.	It takes more calories to eat a piece of celery than the celery has in it to begin with.
	--
35.	Common Cobra venom is not on the list of top 10 venoms yet it is still 40 times more toxic than cyanide.
	--
36.	The human feet perspire half a pint of fluid a day
	--
37.	Julius Caesar's autograph is worth about $2,000,000.
	--
38.	Jim Henson first coined the word "Muppet". It is a combination of "marionette" and "puppet."
	--
39.	An average human loses about 200 head hairs per day.
	--
40.	The state of Florida is bigger than England.
	--

CRYPTOGRAM

1.

G	N	U	S	R		G	R	J	J	V	P		U	N	S	S	P

I	N	W	R		U	R	R	J		F	J	M	E	J		G	M

G	D	N	W	R	S		M	H	H		G	I	R		Q	N	B	B	S	R

N	G		P	Q	R	R	B	P		Z	Q		G	M		1	6	0

F	O	/	I	D	.

2.

H	T	M	I	H		L	R	Q	O		N	D	M	F

G	D	H	O	H	.

CRYPTOGRAM

3.

B J B L Q   W B L H P G ,

V G A O K Y V G M   V Y B G F V A U O

F I V G H ,   Z U H   U   K G V D K B

B Q B   &   F P G M K B   W L V G F

U O P G M   I V F Z   F Z B V L

T V G M B L W L V G F .

--

CRYPTOGRAM

4.

KVI GIJMRERKPDL DH

PLGIFILGILJI NRZ

NEPKKIL DL VIQF

(QREPUCRLR) FRFIE .

--

5.

VWM SQKTZAJ URG VWM

GAGVMF GWAZ SO VWM

VAVRLAJ , RLH GWM

ZFSCAHMH VUMLVK - OACM

KMRFG SO GMFCAJM .

--

CRYPTOGRAM

6.

C		A	P	Y	A	T		R	C	B	,		I	C	B

T	S	B	P	C		P	C	R	E	C	,		T	S	Q	W	O

L	T	Y		D	Y	A	S	D	W		H	S	D

W	D	K	B	F	K	B	X		S	B	Y		Q	K	L	Y	D		S	H

G	Y	Y	D		K	B		4	.	1	1		O	Y	A	S	B	W	O	.

--

CRYPTOGRAM

7.

F J R S   A B J X K Y G   U S J I R Y M A

K   L M D   V K F M S   J Q   X R Y R A

M O M S F   B D J   D M M H A   A J

B G K B   Z B   I J M A L ' B

I Z P M A B   Z B A M V Q .

--

8.

I C W A Q D J F P Y   E P Z P W K   Q D

K D J Z P D L   F P B   S R K U   B Q V

F C D L U K L   U S S Y B .

--

CRYPTOGRAM

9.

| S | B | P | | O | P | D | N | G | | N | A | | S | B | P |

| M | Z | L | S | C | M | J | F | M | D | | X | C | N | K | D |

| L | D | M | V | P | | F | L | | L | N | | H | N | K | P | C | A | Z | J |

| N | D | J | I | | 1 | / | 1 | 4 | , | 0 | 0 | 0 | S | B | | N | A | | M | D |

| N | Z | D | Q | P | | F | L | | P | D | N | Z | Y | B | | S | N |

| V | F | J | J | | M | | B | Z | G | M | D | . |

--

10.

| Z | A | C | | A | N | W | O | E | | I | H | O | U | E | | U | V |

| O | I | G | N | Z | | 7 | 5 | % | | D | O | Z | C | H | . |

--

CRYPTOGRAM

11.

| N | F | U | | R | Q | K | U | H | | J | R | H | H | T | | W | Q |

| Y | U | Q | U | I | E | U | H | R | | R | V | U | | Q | U | R | V | H | A |

| 2 | 0 | | N | W | B | U | T | | N | R | H | H | U | V | | N | F | R | Q |

| Q | W | R | K | R | V | R | | J | R | H | H | T | . |

--

12.

| R | | P | Q | W | H | K | O | Y | N | | J | U | Y | Y |

| E | K | I | W | E | I | U | X | R | X | K | | U | X | H | | O | Q | Q | C |

| R | F | C | | K | R | X | | U | X | | R | I | R | U | F | . |

--

CRYPTOGRAM

13.

BFY LFZSBYLB OSWBWLF

UZDVSGF NVL GFVSEYL

W, NFZ NVL 4 XYYB 9

WDGFYL.

--

CRYPTOGRAM

14.

T RIDZOE'R HOG DR QFJ

T WFYO, GPJ ETJWOET

JETI XFE DJR XFFZ.

JWON TEO TR

DQZDMDZPTK TR

RQFHXKTCOR, HDJW QF

JHF OMOE GODQB JWO

RTYO. RFYO JEFIDATK

RIDZOER WTMO GPDKJ

HOGR FMOE ODBWJOOQ

XOOJ TAEFRR.

CRYPTOGRAM

15.

| Q | F | O | X | Q | U | O | | N | D | C | O | | Y | V | Q | L | | H | C |

| Q | | T | Q | P | H | X | | N | O | Q | U | R | O |

| I | Q | Y | O | I | Q | N | N | : | | 7 | | V | D | W | G | B | O | Y | . |

--

16.

| L | M | Y | X | P | K | | V | D | | K | U | Y | | Q | O | Z | G |

| Y | O | B | Z | V | D | U | | A | Q | M | L | | K | U | X | K |

| Y | O | L | D | | V | O | | K | U | Y | | Z | Y | K | K | Y | M | D |

| " | P | K | . | " |

--

CRYPTOGRAM

17.

MQK LIM IOKF MQK

SKMMKF "G" GW PDSSKL

D MGMMSK.

--

18.

LO PRDY, LR LG

LMMNEDM RU GQNDJ LO

BJUOR UB D ANDA

ZNJGUO.

--

CRYPTOGRAM

19.

WDT KPXT QTTG AZXTI

HLZX " VG " , WDT PLXM

PCCLTBUPWUZK HZL

VTKTLPY GFLGZIT.

--

20.

CW 1933, SCKVLH

SQFTL, OW OWCSOPLA

KOYPQQW KZOYOKPLY,

YLKLCBLA 800,000 NOW

MLPPLYT.

--

CRYPTOGRAM

21.

Z	U	Y	V	M	U	P		M	V	Y	D		V	G		Y	T	F

G	C	U	X	X	F	G	Y		M	K	Q	P	Y	A	D		V	P

Y	T	F		L	K	A	X	W	,		L	V	Y	T		U

O	K	O	Q	X	U	Y	V	K	P		K	B		1	0	0	0

U	P	W		E	Q	G	Y		1	0	8	.	7

U	M	A	F	G	.

CRYPTOGRAM

22.

QKONRIX RF NOXLN RI

ZLBLXKLS LF NUIX LF

PUJA ZLBJROF LBO

BOXRFJOBOQ PNUUQ

QUIUBF.

23.

TQTZL XFJT LCS HFUO

N WXNJD, LCS'ZT

UCVWSJFVK 1/10 CEN

UNHCZFT.

CRYPTOGRAM

24. X L V A   L P Z L R A   V I K C

P B O V   Z N B C   B D E V E C Q L

F L H B .

25. P F C   Q F B H V D C Q V W

T C C P R C   B W   P F C

W P Q V H S C W P   J H B Z J R   J H M

B W   D J N J T R C   V E

R B E P B H S   8 5 0   P B Z C W

B P W   V L H   L C B S F P .

CRYPTOGRAM

26. SZU VMGODUNS KMKU UHUF BJN 11 VUJFN MPX.

27. YKTGHQM YJZ Y TZTIZJ GW VEZ BZYPE WYTRKA.

28. ALFENIFYONIXPBMFI JOIHE DOIL BD APO HQJMOL 13.

CRYPTOGRAM

29.

V I G W Y L B Q 1 , 0 0 0

M T P P W G W Q Y J B Q O C B O W N

B G W N U I F W Q I Q Y L W

K I Q Y T Q W Q Y I P B P G T K B .
--

30.

Y R B J S B K L C D Z R N E H K

U H X E R D P Z N C R N B G L D

L C J N U H X E X B P L D Z L D C

R F S X D N B .
--

CRYPTOGRAM

31.

G K P G P X Q O F   G Q H   L S X X

I P J Z ;   S O   I S M F G O X R

Q V V F G O Z   O K F S M   K F Q M O

Q H I   H F M B P C Z   Z R Z O F Y .

32.

N C   X P W   C Q Q S   L   F Q L D W I I

L I R L - F Q I M U Q J ,   N M F

F M P A L B O   Z N I I   Q G T I P S Q .

33.

K X O E   A F W O E F Y D

Y X H E I F H O   U F O L

O Y I A N O .

CRYPTOGRAM

34.

ZGFR TO ZRFSZDRU

RBRZN OTQ JTYEDRO TY

DXR E.O.

--

35.

GOUHU XHU 45 FVMUK

SJ AUHEUK VA GOU

KWVA SJ X OCFXA

NUVAP.

--

CRYPTOGRAM

36.

Y	C		G	D	M	S	G	I	M	,	1	2

C	M	K	A	Y	S	C	U		K	J	Z	Z		A	M

I	J	D	M	C		E	Y		E	T	M		K	S	Y	C	I

R	G	S	M	C	E	U		M	D	M	S	F		V	G	F	.

--

37.

L	V	C		R	.	Y	.		F	T	R	M	V	L

W	U	W	Y	P	W		S	T	A		2		B	C	G	L	Y

W	G		W	B	A	C		S	A	T	H		A	R	Y	Y	O	W	.

--

CRYPTOGRAM

38.

L S U D U Z D U R T U D 5 8

K P W W P R H Y R O J P H L S U

E J

39.

H G C L R Q T D Y X P V I E P K J R

J T K C W F S P I L B H H .

40.

R Y U S G X S R K D G R P

S W W Q W P L B T Q X D I U U Y A

D I 1 8 6 9 F M S

W Q I P D R P , X D E E D S A

R Q A H E Q .

CRYPTOGRAM

41.

D T U M K B ,   L M P T C   O C   L J M

A . C . ' C   S T U L J M C L

O Q D T Q K   F B U L .

--

42.

B   M D S F   D S   I L X Z L X U L

Z N B Z   Y I   Z N L   I B A L

J S D X Z   B X F   O B U C

(S B U L U B S ,   C B H B C)   Y I

U B G G L F   B

" K B G Y X F S D A L " .

--

CRYPTOGRAM

43.

R	M	T		Y	S	C	Z	U	'	B		S	Z	U	T	B	R

J	G	T	P	T		S	X		P	M	T	Y	G	F	H		H	W	N

G	B		9	0	0	0		K	T	I	C	B		S	Z	U	!

--

44.

B		F	A	Q	I	L	B	L	H		I	S	W	U	N	O	X	V

O	X		T	B	W	W	Q	U		B		H	K	O	H	.

--

45.

N	S	B	L	I	H		D	U		H	Y	B		M	E	X	Q

B	E	O	X	D	U	Y		F	M	S	N		H	Y	L	H

B	E	N	U		D	E		H	Y	B		X	B	H	H	B	S	U

"	I	H	"	.

--

CRYPTOGRAM

46.

N P Y Q P M ,   L T S T M B N T

S B J R   L S B F C   E T   E Z P

F Y Q P Y E F T Y   T U   E Z P

L Z P P R P X A M D P M .

--

47.

1 4 %   Y D   M N N   D M H S R   M G J

R S M S X R S X H R   M F P   V M J P

B U   M G J   2 7 %   Y D   U P Y U N P

Q G Y K   S I M S   D M H S .

--

CRYPTOGRAM

48.

J F "Q J P M F X M W D O V M

P T B K Q ", V T F F J K T P

P M X O W E (T F O V W F N

V W Y I J F Q) F M Z M E

K P J F I Q .

--

CRYPTOGRAM

49.

L D S K | W R D O J V X | X J W | R D

B K D N S V A | J | C A Y | P J L A K

D Z | O S V S W | A M A K L | R Y D

Y A A I W | D K | T R | Y T P P

N T F A W R | T R W A P Z .

--

50.

L Y Q | D R W F L | C N B A U V X B ' F

W Q F L V O W V U L | R U | N V U V B V

G V F | R U | W R N Y C A U B ,

M W R L R F Y | N A X O C M R V .

--

CRYPTOGRAM

51.

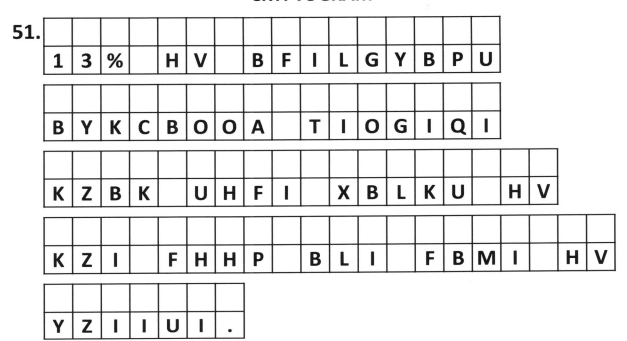

```
1 3 %   H V   B F I L G Y B P U
```

```
B Y K C B O O A   T I O G I Q I
```

```
K Z B K   U H F I   X B L K U   H V
```

```
K Z I   F H H P   B L I   F B M I   H V
```

```
Y Z I I U I .
```

--

52.

```
P I W C L U   I U S   C   O H O C Z
```

```
H M   7 2   Y Q M M S B S L O
```

```
W I U T Z S U   Q L   U F S S T P .
```

--

CRYPTOGRAM

53.

ASDW GP ZJX NTHK

GPHWTF GT ZJX

AWCGDDXWT ZN JWEX W

CWGHCNWF.

--

54.

PYDED JF Z 1 JT 4

AYZTAD PYZP TDM VIEW

MJKK YZND Z MYJPD

AYEJFPCZF.

--

CRYPTOGRAM

55.

P F O I X E , P I P D

P J J V L X J I R D K V X M P F

Q M R F I R Q X W V X W

Y V M D R F .

--

56.

Y C D V ' Z I X S F H Q G N V

Z S R F I C M Q C Y T Q X H O D .

--

57.

J I G Z P R N Y G E R D G Y V Z P

J Z A G E D G S I G E F Q E V

A G J V N Y Z V V .

--

CRYPTOGRAM

58.

E	B	Y	N		W	B	M	F		D	B	K	N		M	D	N

D	S	P	D	N	F	M		A	B	M	N		L	I

D	L	E	L	F	N	H	R	B	Y	S	M	J		L	I		B	O	J

E	B	E	E	B	Y	.

--

CRYPTOGRAM

59.

GWL BWTXTBBTSLO WHO

HDURG 7,100 TOXHSIO,

UK CWTJW USXQ HDURG

460 HYL EUYL GWHS 1

OFRHYL ETXL TS HYLH.

--

60.

EIGGXQ DXISSO IQV

HLJGZXO RUZIJQX IGX

DGTSLXG IQV HJHSXG.

--

CRYPTOGRAM

61.

ADJ SBHJ YX RQVUJBR

FYQAUSJ RQ WJVARPRJO

TYQDJV .

--

62.

GC CPZ KREI FXWXP ,

GX GF GUUPSWU XR JBX

WCK WUVRTRU RC

FBCNWKF JPDREP CRRC .

--

CRYPTOGRAM

63.

J	K J O	R U I U	Q O U	

V W Q U E I H U	O F	

Y I O I H L Q C I	Q S	J	U G J K I	

Q U	O F F	U L J B B	O F

U Z R I I P I	O W H F R A W .

--

64.

V F W	M W U M C W	U Z	N R I G W C

Q U H R O J W	J U I W	V O I P W L R

M W I	Q G M N V G	V F G H	G H L

U V F W I	Q U O H V I L .

--

CRYPTOGRAM

65.

VNBKX KEQIOI SOHSZO

ZNJO, HQ EJOVEBO,

QNQO COEVG ZHQBOV

XKEQ ZOPX KEQIOI

SOHSZO IH.

66.

CDH DIAXF VYOS QE

RYATPQEHO YU 80%

MXCHP.

CRYPTOGRAM

67.

RMPUF SSZWK QMENJB'C

CZU EULYUBYK SEUTZSU

SZNC.

68.

BOU HIUXHVU SUHE

RUZLCS PCSS EXHP H

SCZU 35 FCSUT SGZV

GX PXCBU

HRRXGNCFHBUSA 50,000

UZVSCTO PGXET.

CRYPTOGRAM

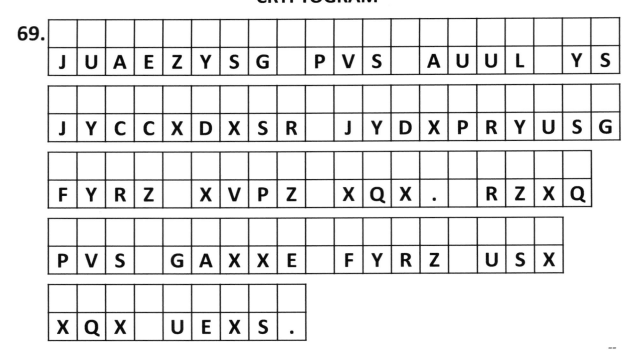

69.

| J | U | A | E | Z | Y | S | G | | P | V | S | | A | U | U | L | | Y | S |

| J | Y | C | C | X | D | X | S | R | | J | Y | D | X | P | R | Y | U | S | G |

| F | Y | R | Z | | X | V | P | Z | | X | Q | X | . | | R | Z | X | Q |

| P | V | S | | G | A | X | X | E | | F | Y | R | Z | | U | S | X |

| X | Q | X | | U | E | X | S | . |

--

70.

| F | E | D | A | | F | E | Z | | U | E | | J | C | F |

| V | U | D | D | U | F | E | | K | Z | F | K | D | Z | | C | U | D | D |

| D | U | P | Z | | J | F | | V | Z | | 1 | 1 | 6 | | F | O |

| F | D | H | Z | O | . |

--

CRYPTOGRAM

71.

QV BRZ DTMB 4000

GZTCM VJ VZP TVQSTDM

RTLZ OZZV

FJSZMBQITBZF .

72.

Y TCUV FYE LDS Y

BPEEVU 300 WVVB (91

T) UCES DE GPHB CEV

EDSOB .

CRYPTOGRAM

73.

U	C		R	E	Y	J	,	G	
'	P	E							

B	T	Z	T	D	'		U	L		G		L	M	E	Z
T		E	J												

1	0	.

CRYPTOGRAM

74.

| Q | Z | U | O | | S | C | P | P | M | H | L | J | B | M | Y | | G | Z | U |

| P | Z | Y | P | M | | I | B | P | X | | P | X | M | B | H |

| L | M | M | P | | P | R | | L | B | U | F | | R | C | P |

| I | X | M | P | X | M | H | | P | X | M | | J | M | Z | L |

| P | X | M | O | | Y | B | P | | R | U | | B | Y | | V | R | R | F |

| P | R | | J | Z | O | | M | V | V | Y | | R | U | | P | R | | S | M |

| P | X | M | B | H | | G | Z | P | M | H | W | B | J | J | Z | H | Y | ' |

| L | R | R | F | | R | H | | U | R | P | . |

--

CRYPTOGRAM

75.

AZG RQNUH'K CQLDPGKA

BXNGDAK RGNG 8 XDH 9

XDH UEWGH ED IZEDX

ED 1910.

--

76.

QXT QBBQXVYMCX OKC

FLETLQTS FL NXFLK FL

1498.

--

CRYPTOGRAM

77.

| G | Y | G | Z | R | C | M | A | F | | T | M | Y | J | | F | Y | G | G | Z |

| L | T | D | | A | U | T | | R | T | V | D | F | | G | C | X | R |

| E | C | J | . |

78.

| U | E | G | N | | E | R | I | | F | Q | U | N | S | N |

| H | E | R | ' | G | | T | Q | A | X | G | . |

79.

| I | X | K | O | H | Y | | X | Y | J | | O | | R | A | R | O | P |

| A | Q | | 7 | 2 | | L | B | Q | Q | J | Z | J | H | R |

| K | X | Y | N | P | J | Y | | B | H | | Y | E | J | J | N | I | . |

CRYPTOGRAM

80.

Y	M	X	I	G	J	P		I	Z	S	S		O	M	E	G

K	G	M	K	S	G		D	N	R	X		K	S	R	X	G

F	E	R	P	N	G	P	.

--

CRYPTOGRAM - HINTS (1/2)

#		#	
1.	R => E, U => B, H => F	27.	J => R, K => L, T => M
2.	O => E, H => S, L => H	28.	E => S, F => I, H => N
3.	U => A, L => R, B => E	29.	W => E, G => R, Q => N
4.	R => A, I => E, L => N	30.	D => T, N => O, H => L
5.	A => I, O => F, F => R	31.	B => V, L => K, R => Y
6.	W => D, Y => E, S => O	32.	Q => E, I => L, A => M
7.	B => T, K => A, O => V	33.	Y => C, E => T, K => M
8.	P => A, Q => I, K => E	34.	S => O, R => E, T => I
9.	K => W, Z => U, S => T	35.	N => B, K => S, A => N
10.	O => A, H => R, Z => T	36.	E => T, F => Y, M => E
11.	H => L, T => S, U => E	37.	A => R, G => N, C => E
12.	R => A, K => E, U => I	38.	R => O, W => L, U => E
13.	V => A, Y => E, F => H	39.	C => I, D => M, P => A
14.	R => S, M => V, O => E	40.	P => T, R => S, A => M
15.	Q => A, U => G, H => O	41.	B => O, C => S, D => L
16.	K => T, V => I, U => H	42.	Y => I, C => K, B => A
17.	S => L, F => R, M => T	43.	M => H, T => E, R => T
18.	Z => P, J => R, R => T	44.	O => I, K => W, S => O
19.	G => P, T => E, A => C	45.	O => G, E => N, Y => H
20.	Y => R, S => M, N => F	46.	T => O, Q => V, Z => H
21.	Y => T, M => C, D => Y	47.	Q => K, P => E, G => N
22.	O => E, F => S, L => A	48.	E => R, M => E, F => N
23.	S => U, X => T, L => Y	49.	Y => W, J => A, L => Y
24.	R => Y, A => S, L => A	50.	R => I, F => S, B => D
25.	T => B, W => S, E => F	51.	L => R, I => E, K => T
26.	S => T, P => L, U => E	52.	M => F, U => S, S => E

53.	T => N, H => L, J => H	**79.**	Y => S, H => N, B => I
54.	I => O, Y => H, M => W	**80.**	O => M, K => P, J => Y
55.	E => H, Q => F, X => A		
56.	Y => C, S => N, I => R		
57.	E => A, J => T, Y => R		
58.	F => S, B => A, L => O		
59.	X => L, H => A, S => N		
60.	S => T, U => C, L => H		
61.	R => I, A => T, Y => O		
62.	C => N, J => B, E => R		
63.	Z => Q, F => O, P => Z		
64.	H => N, Q => C, F => H		
65.	I => D, J => V, O => E		
66.	M => W, H => E, Q => I		
67.	P => A, S => L, B => T		
68.	S => L, U => E, G => O		
69.	P => C, R => T, X => E		
70.	J => T, F => O, Z => E		
71.	Z => E, R => H, S => M		
72.	C => O, B => T, V => E		
73.	L => S, Z => R, E => O		
74.	M => E, Z => A, R => O		
75.	A => T, W => V, L => U		
76.	L => N, X => H, Y => R		
77.	C => A, D => R, G => E		
78.	Q => O, I => D, G => T		

1.	Table tennis balls have been known to travel off the paddle at speeds up to 160 km/hr. --
2.	Slugs have four noses. --
3.	Every person, including identical twins, has a unique eye & tongue print along with their fingerprint. --
4.	The Declaration of Independence was written on hemp (marijuana) paper. --
5.	The Olympic was the sister ship of the Titanic, and she provided twenty-five years of service. --
6.	A Czech man, Jan Honza Zampa, holds the record for drinking one liter of beer in 4.11 seconds. --
7.	Your stomach produces a new layer of mucus every two weeks so that it doesn't digest itself. --
8.	Buckingham Palace in England has over six hundred rooms. --
9.	The venom of the Australian Brown Snake is so powerful only 1/14,000th of an ounce is enough to kill a human. --
10.	The human brain is about 75% water. --
11.	The angel falls in Venezuela are nearly 20 times taller than Niagara Falls. --
12.	A housefly will regurgitate its food and eat it again. --
13.	The shortest British monarch was Charles I, who was 4 feet 9 inches. --
14.	A spider's web is not a home, but rather a trap for its food. They are as individual as snowflakes, with no two ever being the same. Some tropical spiders have built webs over eighteen feet across. --

CRYPTOGRAM - ANSWERS (2/6)

15.	Average life span of a major league baseball: 7 pitches.
	--
16.	Dreamt is the only English word that ends in the letters "mt."
	--
17.	The dot over the letter "i" is called a tittle.
	--
18.	In Utah, it is illegal to swear in front of a dead person.
	--
19.	The name Jeep comes from "GP", the army abbreviation for General Purpose.
	--
20.	In 1933, Mickey Mouse, an animated cartoon character, received 800,000 fan letters.
	--
21.	Vatican City is the smallest country in the world, with a population of 1000 and just 108.7 acres.
	--
22.	Dueling is legal in Paraguay as long as both parties are registered blood donors.
	--
23.	Every time you lick a stamp, you're consuming 1/10 of a calorie.
	--
24.	Bats always turn left when exiting a cave.
	--
25.	The rhinoceros beetle is the strongest animal and is capable of lifting 850 times its own weight.
	--
26.	The youngest pope ever was 11 years old.
	--
27.	Almonds are a member of the peach family.
	--
28.	Triskaidekaphobia means fear of the number 13.
	--
29.	More than 1,000 different languages are spoken on the continent of Africa.
	--

30.	Mercury is the only planet whose orbit is coplanar with its equator.	--
31.	Chocolate can kill dogs; it directly affects their heart and nervous system.	--
32.	If you feed a seagull Alka-Seltzer, its stomach will explode.	--
33.	Most lipstick contains fish scales.	--
34.	Rape is reported every six minutes in the U.S.	--
35.	There are 45 miles of nerves in the skin of a human being.	--
36.	On average, 12 newborns will be given to the wrong parents every day.	--
37.	The U.S. bought Alaska for 2 cents an acre from Russia.	--
38.	There are over 58 million dogs in the US	--
39.	Flies jump backwards during takeoff.	--
40.	Sugar was first added to chewing gum in 1869 by a dentist, William Semple.	--
41.	Laredo, Texas is the U.S.'s farthest inland port.	--
42.	A word or sentence that is the same front and back (racecar, kayak) is called a "palindrome".	--
43.	The world's oldest piece of chewing gum is 9000 years old!	--
44.	A pregnant goldfish is called a twit.	--

CRYPTOGRAM - ANSWERS (4/6)

45.	Dreamt is the only English word that ends in the letters "MT". --
46.	Denver, Colorado lays claim to the invention of the cheeseburger. --
47.	14% of all facts and statistics are made up and 27% of people know that fact. --
48.	In "Silence of the Lambs", Hannibal Lector (Anthony Hopkins) never blinks. --
49.	Your stomach has to produce a new layer of mucus every two weeks or it will digest itself. --
50.	The first McDonald's restaurant in Canada was in Richmond, British Columbia. --
51.	13% of Americans actually believe that some parts of the moon are made of cheese. --
52.	Humans use a total of 72 different muscles in speech. --
53.	Cuba is the only island in the Caribbean to have a railroad. --
54.	There is a 1 in 4 chance that New York will have a white Christmas. --
55.	In Utah, it is illegal to swear in front of a dead person. --
56.	Cats' urine glows under a black light. --
57.	The only real person to be a PEZ head was Betsy Ross. --
58.	Male bats have the highest rate of homosexuality of any mammal. --

CRYPTOGRAM - ANSWERS (5/6)

59.	The Philippines has about 7,100 islands, of which only about 460 are more than 1 square mile in area. --
60.	Warren Beatty and Shirley McLaine are brother and sister. --
61.	The glue on Israeli postage is certified kosher. --
62.	In New York State, it is illegal to but any alcohol on Sundays before noon. --
63.	A cat uses its whiskers to determine if a space is too small to squeeze through. --
64.	The people of Israel consume more turkeys per capita than any other country. --
65.	Right handed people live, on average, nine years longer than left handed people do. --
66.	The human body is comprised of 80% water. --
67.	Frank Lloyd Wright's son invented Lincoln Logs. --
68.	The average lead pencil will draw a line 35 miles long or write approximately 50,000 English words. --
69.	Dolphins can look in different directions with each eye. They can sleep with one eye open. --
70.	Only one in two billion people will live to be 116 or older. --
71.	In the last 4000 years no new animals have been domesticated. --
72.	A mole can dig a tunnel 300 feet (91 m) long in just one night. --
73.	In golf, a 'Bo Derek' is a score of 10. --

CRYPTOGRAM - ANSWERS (6/6)

74.	Many butterflies can taste with their feet to find out whether the leaf they sit on is good to lay eggs on to be their caterpillars' food or not. --
75.	The world's youngest parents were 8 and 9 and lived in China in 1910. --
76.	The toothbrush was invented in China in 1498. --
77.	Elephants only sleep for two hours each day. --
78.	Rats and horses can't vomit. --
79.	Humans use a total of 72 different muscles in speech. --
80.	Donkeys kill more people than plane crashes. --

Printed in Great Britain
by Amazon

15055103R00070